# WHY IT'S NOT CARBON DIOXIDE AFTER ALL

### Douglas J Cotton

The opinions expressed in this manuscript are solely the opinions of the author and do not represent the opinions or thoughts of the publisher. The author has represented and warranted full ownership and/or legal right to publish all the materials in this book.

Why It's Not Carbon Dioxide After All
All Rights Reserved.
Copyright © 2014 Douglas J Cotton
v1.0

Cover Photo © 2014 Douglas J Cotton. All rights reserved - used with permission.

This book may not be reproduced, transmitted, or stored in whole or in part by any means, including graphic, electronic, or mechanical without the express written consent of the publisher except in the case of brief quotations embodied in critical articles and reviews.

Outskirts Press, Inc.
http://www.outskirtspress.com

ISBN: 978-1-4787-2922-8

Outskirts Press and the "OP" logo are trademarks belonging to Outskirts Press, Inc.

PRINTED IN THE UNITED STATES OF AMERICA

# Table of Contents

**Chapter 1:** Introduction ...................................................................................1

**Chapter 2:** A Slice of History............................................................................4

**Chapter 3:** Physics: the Misunderstood Science .........................................8

**Chapter 4:** Where It All Went Horribly Wrong .......................................14

**Chapter 5:** It's All About Restoring Equilibrium ......................................17

**Chapter 6:** Heat Creep Provides the Answer ............................................22

**Chapter 7:** So Why Is the Globe Warming? ..............................................27

**Appendix:** Temperature-Rainfall Correlation ...........................................29

# 1
## Introduction

The reader will be very much aware of the extent to which the concept has been propagated that carbon dioxide supposedly causes a "greenhouse effect" which is to blame for climate change, and that mankind should thus endeavour to reduce emissions of carbon dioxide into the atmosphere.

The contribution by humans is thought to have caused "anthropogenic global warming" (AGW) because carbon dioxide levels have been rising and so too did temperatures, especially in the 30 years or so leading up to the maximum in 1998. Even though the world is now in the middle of about 30 years of very slight overall cooling, whilst carbon dioxide levels continue to rise, the greenhouse conjecture gained so much momentum late last century as to be almost beyond quashing for several years yet.

Essentially the concept dates back to somewhat primitive experiments conducted in the nineteenth century, but soundly rebutted in the mid twentieth century by several eminent scientists who explained that carbon dioxide and water vapour could do nothing but cause cooling. However, from around 1980 a relatively small group of climatologists rekindled the idea, supposedly supporting it with what amounts to totally invalid and misunderstood physics. Politicians and most of the media have been convinced, and the US Government alone now invests over $22 billion dollars a year into funding climatology research and taking action to reduce so-called "carbon" pollution. Many more billions will be spent by other governments and by consumers forced to pay higher energy bills.

In fact carbon dioxide is a tasteless, colourless and odourless gas comprising about 0.04% of Earth's atmosphere, whilst most of our air is made up of about four-fifths nitrogen and one-fifth oxygen. The level of water vapour varies between about 1% and 4% and so it is by far the most dominant "greenhouse gas" with about 25 to 100 times as many molecules as there are carbon dioxide molecules. It is a relatively simple matter to show, as in the Appendix, that regions become cooler as the level of water vapour increases, but climatologists would have us believe the exact opposite.

It is not disputed for one moment that we should endeavour to reduce real pollution which contributes to the smog that presents serious health hazards, especially in some

Asian cities. But carbon dioxide is not smog and it is absolutely essential for all plant life as it combines with water and energy from the Sun in a photosynthesis process. That is how the wood in the trees and the coal in the ground come to contain carbon. Plant life will thrive and the world will be able to produce more food if carbon dioxide levels continue to rise. Indeed there have been times in the distant past when such levels have been far higher, and there is some evidence that these levels are a result, not a cause of temperature cycles.

Many scientists and academics have started to realise that there are serious errors in the greenhouse conjecture, and organisations such as *Principia Scientific International (PSI)* and *The Heartland Institute* have attracted hundreds of members. However, when it comes to the detail as to just exactly what processes are at play in determining planetary atmospheric and surface temperatures, these organisations have not been able to produce a sound physical explanation which gels with observed facts. For example, the author has asked several key members of *PSI* if they can explain how the necessary energy gets to the surface of Venus, the deep atmosphere of Uranus or even the core of our own Moon in order to retain the temperatures that exist therein. In the author's opinion, no valid explanation has been forthcoming, and yet these are critical issues very relevant to what is happening on Earth, because the laws and processes in physics apply throughout the universe.

This book purposefully does not venture into discussion as to whether or not there is a correlation between temperatures and mankind's contribution to carbon dioxide levels in the atmosphere. Nor is it disputed that there is indeed still some long-term global warming. The final chapter attributes such warming as being entirely due to several natural and overlapping climate cycles, both short and long term, and carbon dioxide levels are irrelevant.

There is, instead, a detailed explanation of the physics involved (hopefully written in a manner that is understandable for those without much background in the subject) and the book puts forward a universal hypothesis as to the processes which are considered to be determining all planetary atmospheric and surface temperatures, and even sub-surface temperatures right down to the core.

*The hypothesis in this book is supported by all known and estimated temperature data throughout our Solar System, whereas the greenhouse conjecture is demolished from various angles and never has been successfully applied to explain any other planetary temperature data.*

# INTRODUCTION

The author throws down the gauntlet to any reader who, after comprehending the arguments herein, believes it possible to rebut same, be it in public or private debate. In particular, those with a sound understanding of thermodynamics are asked for their opinions on the content. It is hoped that some will stand in the public arena and join the rising protests against the pseudo physics that has been promulgated by the *Intergovernmental Panel on Climate Change (IPCC)* in Switzerland whose authors have had an alarming influence, whilst attracting huge sums of money to the cause. One wonders how many lives may have been saved had such funds been devoted instead to humanitarian aid.

# 2
## A Slice of History

The world will one day look back upon a small slice of history that began in the 1980's and sadly have to conclude that never in the name of science have so many people been so seriously misled by so few for so long. Never have so many careers, so much time and so much money been spent in the pursuit of such a misguided and ineffective goal to reduce human emissions of carbon dioxide, which comprises about one molecule in every two and a half thousand other molecules in the atmosphere of our planet, Earth.

The author is very much aware of the arguments put forward and the extent of vested interests dependent upon literally billions of dollars of government funds shelled out in the belief that "the science is settled" and now we must get on with the task of "saving the planet" by cutting "carbon" emissions no matter what the cost to society. He is also aware from personal experience in debate with many hundreds of believers and so-called "deniers" that very, very few exhibit a valid understanding of the relevant physics and physical laws in the field of thermodynamics.

The physics involved is actually at the forefront of current knowledge, and the hypothesis put forward in this book has not, to the author's knowledge, been published anywhere else in world literature. Yet the IPCC authors make it sound all too obvious, and even young school children are taught in simplistic terms by misled teachers that this wicked "pollutant" carbon dioxide acts like a blanket making the world a hotter place. It is incorrectly accused of allowing all incoming radiation from the Sun to strike the Earth's surface, but "trapping" all upwelling long-wave infrared radiation. Then it is assumed that, when you warm up some very cold molecules of air in the atmosphere by a fraction of a degree, that extra energy somehow returns to Earth and makes it hotter, like hot air physically trapped in a glass greenhouse.

In seemingly more sophisticated arguments the Intergovernmental Panel on Climate Change puts forward the following "explanation" in the Glossary on their website www.ipcc.ch where we read: *"Greenhouse gases effectively absorb thermal infrared radiation, emitted by the Earth's surface, by the atmosphere itself due to the same gases, and by clouds. Atmospheric radiation is emitted to all sides, including downward to the Earth's surface. Thus, greenhouse gases trap heat within the surface-troposphere system. This is called the greenhouse effect."*

# A SLICE OF HISTORY

The first problem we strike with this IPCC "explanation" is that radiation from a cooler atmosphere cannot actually transfer thermal energy to a warmer surface. Radiation does not work that way. This is a complex area of physics not well understood, but explained in far more detail in the author's paper *"Radiated Energy and the Second Law of Thermodynamics"* published on several websites in March 2012. [1 & 2]

But before we continue with the IPCC description of the "Greenhouse Effect" let us turn to their summary of which gases are "Greenhouse Gases."

*"Greenhouse gases are those gaseous constituents of the atmosphere, both natural and anthropogenic, which absorb and emit radiation at specific wavelengths within the spectrum of thermal infrared radiation emitted by the Earth's surface, by the atmosphere itself, and by clouds. This property causes the greenhouse effect. Water vapor (H2O), carbon dioxide (CO2), nitrous oxide (N2O), methane (CH4), and ozone (O3) are the primary greenhouse gases in the Earth's atmosphere."*

They go on to list some other such gases that are present in far smaller proportions, but they do not in any way emphasise the fact that water vapour is by far the most prolific among these so-called greenhouse gases. There is, none-the-less, a clear implication that all such gases act to warm the surface of Earth and presumably other planets. But there are far more water vapour molecules and water droplets and so we would expect these to have the most significant warming effect. The problem is that actual temperature data (such as that in the author's study in the Appendix) shows that water vapour has an overall cooling effect, reducing both the mean daily maximum and minimum temperatures in more moist regions. It would take relatively little to fund a more comprehensive study of actual temperature data in moist and dry regions and such a study would almost certainly confirm the conclusions in the study in the Appendix. This easily analysed and confirmed fact should in itself be sufficient to demolish the "science" that alleges a greenhouse effect is causing global warming.

Continuing with the first quote above, the IPCC puts it this way ...

*"Thermal infrared radiation in the troposphere is strongly coupled to the temperature of the atmosphere at the altitude at which it is emitted. In the troposphere, the temperature generally decreases with height. Effectively, infrared radiation emitted to space originates from an altitude with a temperature of, on average, -19°C, in balance with the net incoming solar radiation, whereas the Earth's surface is kept at a much higher temperature of, on average, 14°C. An increase in the concentration of greenhouse gases leads to an increased infrared opacity of the atmosphere and therefore to an effective radiation into space from a higher*

*altitude at a lower temperature. This causes a radiative forcing that leads to an enhancement of the greenhouse effect, the so-called enhanced greenhouse effect."*

Now, the overall level of temperatures in the Earth's atmosphere is determined by a tendency for the whole system to radiate back to space approximately the same amount of radiative flux that is received from the Sun. This radiation to space may be perhaps up to 0.5% less than what is received during a long period of natural warming, and likewise up to 0.5% more when the world is cooling as it did, for example, between the Medieval Warming Period (about 900 years ago) and the Little Ice Age about 400 years ago. Attempts to measure any difference between incoming and outgoing radiation at the top of the atmosphere (TOA) are not accurate enough to determine which is the greater, and such measurements would need to be done simultaneously all over the globe. The fact that wavelengths that are absorbed by carbon dioxide are missing proves nothing because the warmed carbon dioxide molecules can pass on energy by diffusion and, as a result, much of that absorbed energy ends up being radiated to space by the more prolific water vapour molecules. So there is no measured evidence of any radiative forcing and, even when there is radiative imbalance at TOA, it is not the *cause* of warming or cooling, but rather the *result* thereof.

When the IPCC authors write *"Effectively, infrared radiation emitted to space originates from an altitude with a temperature of, on average, -19°C ..."* they are assuming that -19°C is the mean temperature of the whole Earth-plus-atmosphere system. Whether or not that is an accurate estimate, they grossly oversimplify the whole system to the extent that all their subsequent conclusions are invalid. The altitude to which they refer is about 5Km, whereas what we will call the "pivoting altitude" (where equal amounts of outward radiative flux come from above and below) is probably between 3Km and 4Km because a significant amount of radiation comes from the Earth's surface directly to space. Also, because radiative flux is not proportional to temperature but instead increases with the fourth power of temperature, and because there is more water and water vapour below the pivoting altitude than above it, that altitude must be lower than that where the temperature is -19°C.. So it is totally incorrect to claim that any meaning can be attached to an altitude where the temperature is -19°C. But, more importantly, they gloss over why any increase in this altitude should lead to a warmer surface, especially when they acknowledge that there would be a lower temperature at a higher altitude.

What does happen is that, as any particular column of the troposphere becomes more moist, the slope of the plot of temperature against altitude becomes less steep, as is well known. This causes the whole temperature plot to rotate around the pivoting altitude, because such rotation ensures that the total outgoing radiative flux continues to match closely the total incoming radiative flux from the Sun. This then leads to lower "supporting" temperatures at the surface as is explained in more detail in a later chapter.

Notice that the IPCC confirms that greenhouse gases both *"absorb and emit"* long wave (low energy) infrared radiation. They are better referred to as "radiating gases" because they actually help thermal energy escape more quickly to space, and so they certainly provide no "blanket" or insulating effect – quite the opposite. They are more like holes in a blanket because they radiate away to space the energy in non-radiating nitrogen and oxygen which was acquired by conduction from the surface and diffusion within the atmosphere. Those who construct double glazed windows in buildings know that if moist air enters the gap between the panes the insulation effect is reduced, not improved. They get the best insulation with the inert (non-radiating) gas argon, but dry air is usually sufficient. Energy moves across the gap at a relatively slow rate by the conduction-like process of diffusion which involves molecules passing on energy in collisions. If we introduce radiating molecules between the panes they will transfer thermal energy at the speed of light, but only ever from warmer groups of radiating molecules to cooler groups of similar molecules. Hence this energy "leap frogs" the thermal energy in the far slower moving molecules that are transferring such energy by diffusion.

In summary, it is not hard to demolish the "pseudo" physics promulgated by IPCC authors, primarily because water vapour is seen to produce lower surface temperatures and carbon dioxide acts in the same way, being just another radiating gas. The hypothesis put forward in this book explains why this is the case, and it may be used to answer some important questions such as "Why is a planet's surface much hotter than direct solar radiation could make it?" and "Why have planets not just cooled off over the course of billions of years?" An understanding of what happens on other planets is vital to an understanding of Earth's climate mechanisms.

# 3
# Physics: the Misunderstood Science

Physics is a universal science, meaning that its laws apply throughout the Universe. When Sir Isaac Newton realised that an apple fell off a tree because of the force of gravity, he was able to extend that "understanding" to explain how the force of gravity keeps the planets in orbit, and that force can be calculated from an equation (or formula) involving the distance between any two bodies and their masses. No doubt many before him had seen apples falling, but Newton *understood* what must be happening.

An understanding of planetary atmospheric and surface temperatures requires an understanding of thermodynamic physics and, in particular, the Second Law of Thermodynamics, which is explained in detail in Chapters 5 and 6. We will be delving into how thermal energy is transferred by radiation and also by non-radiative processes, and we will be looking for universality by checking that our hypothesis works for known temperature data from other planets, notably Venus and Uranus. Venus has some similarities with Earth, whereas Uranus is a good example of a gaseous planet without the complications of ongoing contraction causing energy generation in the centre. The fact that Uranus does not have any significant imbalance in radiative flux implies that it is probably correct to assume that it does have a solid core about half the mass of Earth which prevents its atmosphere collapsing. That atmosphere (predominantly hydrogen plus about 15% helium) reaches many thousands of kilometres above the surface of the Uranus core, and we will see that it is the height of the atmosphere which contributes to the very hot temperatures in the core.

We have already mentioned that the concept of "radiative forcing" referred to in IPCC documentation clearly does not function on other planets such as Venus, where even all the solar energy reaching the top of its atmosphere would be far less than would be required to maintain its surface temperature solely by radiation.

How have so many scientists been so misled by this conjecture that radiative forcing supposedly warms a planet's surface well beyond any temperature that direct Solar radiation could achieve? Perhaps it is because of a growing mentality that all we need is a bit of First Year university physics to grab a formula and that, without understanding the limitations and prerequisites, we can just plug values into that formula and get right answers.

At the centre of all the pseudo "science" associated with the greenhouse conjecture is just such a misapplication of the Stefan Boltzmann Law (SBL) which relates to so-called blackbodies and the temperatures they can reach when subjected to a given flux of radiation.

Let us then seek a better understanding of the SBL and just what it tells us, and does not tell us. Firstly, it is important to understand that the law only applies to true black and grey bodies. The analogy with colours probably stems from the fact that a black object absorbs most of the visible portion (the light) in the Sun's radiation, whereas white objects reflect most colours in the visible spectrum, and grey objects fall in between. But more importantly, we need to understand that the SBL assumes that these bodies gain and lose thermal energy *only* by radiation. This basically restricts them to being objects in outer space where there is no loss of energy by non-radiative processes. The law simply *cannot* be applied to the Earth's surface because that surface is simultaneously losing (and maybe gaining) other energy by non-radiative processes, so that the Sun's direct radiation cannot raise it to the observed mean temperature.

Spontaneous radiation from a blackbody exhibits a wide range of wavelengths with a distribution that is similar in shape for almost any temperature. For example, the first graphic below illustrates the distribution of wavelengths in Solar radiation and you can see that nearly half of the solar radiation is in the infrared band. The second graphic shows some other distributions for much cooler bodies plotted against the frequencies. The Nobel Prize winning scientist Max Planck documented these distributions mathematically, and Wilhelm Wien showed that the peak frequency is proportional to the absolute (K) temperature, as is illustrated in the second graphic.

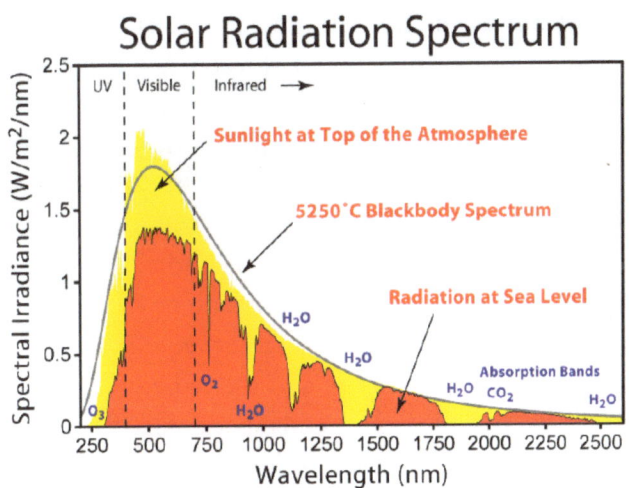

The above graphic from the author's paper [1 & 2] shows that about a fifth of the Sun's incoming radiation is absorbed by the atmosphere. Most of this absorption is by ozone and water vapour, but *some Solar radiation is also absorbed by carbon dioxide* in the vicinity of 2000nm wavelengths. In addition, about 30% of incoming radiation is reflected back to space, and so about half reaches the Earth's surface.

In the case of Venus (with an atmosphere over 90 times the mass of Earth's atmosphere) only about 2% to 3% gets through to its surface, and it should be obvious that this does not raise the Venus surface temperature by hundreds of degrees. The atmosphere on Venus is mostly carbon dioxide and, during the Venus daytime, that carbon dioxide is absorbing over 97% of the incident solar radiation as it enters the Venus atmosphere. Then, during the Venus night, the carbon dioxide is radiating back to space an equivalent amount of energy.

As physics is universal, the evidence on Venus indicates that the radiative forcing concept and the associated greenhouse conjecture are both totally false. These concepts were based on an incorrect understanding of the Second Law of Thermodynamics, which tells us that radiation transfers thermal energy only from warmer to cooler regions of a planet's atmosphere and thence to space, not to a warmer surface.

Thermal energy is represented by kinetic energy (the energy of motion) in the molecules of any substance, be it solid, liquid or gas. But radiation can pass through the vacuum of space and so there are no molecules involved and thus no thermal energy – only electromagnetic energy. If a blackbody is subjected to a steady radiative flux it reaches an equilibrium temperature at which it is radiating back to space the same amount of radiation that it is receiving. The electromagnetic energy in the radiation is not being converted to thermal energy, and nor is thermal energy in the blackbody being converted to electromagnetic energy. What actually happens is that the incident radiation is immediately re-emitted and the effect looks similar to a form of random scattering of the incident radiation. If the source of the incident radiation is hotter than the target and the intensity of the radiation increases, then some of the extra electromagnetic energy in the radiation will be converted to thermal energy in the target, thus raising its temperature until a new equilibrium is attained. The relationship between the absolute temperature and the radiative flux is expressed in the Stefan-Boltzmann Law, the main thing to remember being that radiative flux increases in proportion to the fourth power of the absolute temperature.

There is a frequency distribution for the radiation from a blackbody and this is determined by the Planck function which takes on the shapes indicated below. The curves for cooler temperatures are always contained within the area under the curves for warmer temperatures. The electromagnetic energy which is actually converted to thermal energy in a cooler target can be represented by the area between the Planck curves for the warmer source and the cooler target. That radiation which is represented by the area under the Planck curve for the cooler target is also common to some of the radiation from the warmer source. It is this common radiation which is immediately re-emitted and the term "pseudo scattered" is sometimes used because the result appears the same as if the incident radiation had been randomly scattered. All radiation from a cooler source (such as the Earth's atmosphere) to a warmer target (such as the surface) is thus "pseudo scattered" and none of its electromagnetic energy is converted to thermal energy. As explained in the author's paper [1 & 2] there is a resonance process which takes place only for that incident radiation which also forms a part of the potential outward radiation from the target. This is how the target "recognises" whether the source of radiation is from a body that is warmer or cooler than itself.

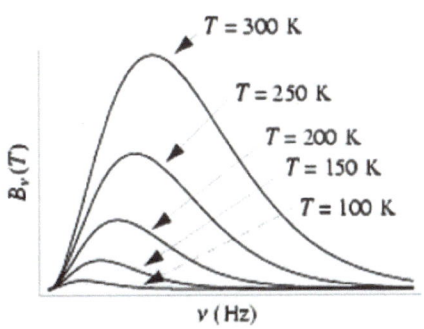

So, as explained above, the radiation that emanates from a blackbody has a frequency distribution, and Wien's Displacement Law tells us that the peak frequency increases proportionally with the temperature. That peak frequency can be seen in the above Planck curves and we should note that the area under the curve represents the total radiative flux for the indicated temperature. But it is important to understand that not all of the electromagnetic energy in the emanating radiation actually came from thermal energy in the body itself. In the case of the Earth's surface, much of the electromagnetic energy in the emanating radiation actually comes from the electromagnetic energy in radiation from the cooler atmosphere. This radiation is immediately re-emitted by the surface *without any of its electromagnetic energy being converted to thermal energy*. The incident radiation from a cooler source does, however, slow the rate of radiative cooling of the warmer target because the target does not have to use as much of its own thermal energy in order to fulfil its "quota" of radiation as is determined by the area under its Planck curve.

So, yes, the so-called "back radiation" does in fact slow down that portion of surface cooling which is itself due to radiation. However, because its energy does not go through the complicated process of being absorbed and converted to thermal energy, the back radiation can have no effect on the rate of non-radiative cooling of any planet's surface. The only radiation that can increase the temperature of the surface must come from a hotter source, namely the Sun. Thus the Solar radiation getting through to a planet's surface is the only radiation that plays a part in determining a planet's surface temperature. Neither on Earth or Venus (or any other planet with a significant atmosphere) does that radiation account for the actual observed planetary surface temperatures, and this fact alone is sufficient to put to rest all the literature and "settled science" which blames global warming on back radiation from the radiating gases in the atmosphere.

The reader might be thinking that, if back radiation slows surface cooling then it leads to warmer mean temperatures. In response we ask, "Cooling from what temperature?" The point is, the whole concept that back radiation is the cause of that "33 degrees of warming" is false, because back radiation does not add to the warming effect of the Sun. We have to ask why the temperature of the atmosphere is what is observed, because that is what determines the level of back radiation, and it is not due to any radiated energy being trapped. As we shall see in a later chapter, it is the pre-determined temperature at the base of the Earth's troposphere which "supports" the surface temperature, because all cooling processes, both radiative and non-radiative, slow down as the temperature gap narrows. This is why the Earth's surface does not continue cooling as quickly in the early pre-dawn hours as it may have done in the late afternoon after a hot sunny morning. If the temperature of the surface were being maintained solely by incident Solar radiation, then we would expect far faster cooling at night, such as is observed on the Moon.

If you have read about a "runaway greenhouse effect" due to all the carbon dioxide in the Venus atmosphere, then you have also been seriously misled. When radiation sets and maintains the temperature of a blackbody that temperature is an absolute maximum for any given radiative flux. The object does not continue to get hotter still, because it is losing energy at the same rate. In practice a planet's surface would need a lot more incident solar radiation than SBL indicates in order to supply the extra energy that is being lost by non-radiative processes. Even if the Venus surface were a true blackbody it would require over 16,000 watts of direct solar radiation striking each square metre of its surface. That is several times what is received at the top of its atmosphere and no amount of back radiation could multiply the energy originally received by a factor of about five. In fact the atmosphere absorbs nearly all of it and, according to measurements made by Russian probes dropped onto the Venus surface, the actual incident solar radiation is less than 20 watts per square metre.

If the carbon dioxide in the Venus atmosphere is still troubling you, look back to the graphic of the Solar Radiation Spectrum above and note the yellow section around 2,000nm marked as being incoming solar radiation absorbed by carbon dioxide. It is carbon dioxide which is absorbing over 97% of solar radiation passing down through the Venus atmosphere. That must surely seem like a cooling effect because it is also radiating that energy back to space before it could ever get to the Venus surface.

So we seem to have a major dilemma which is very obvious on Venus, but just as much an issue on Earth also. How does the required energy get into planetary surfaces? If it is not by radiative processes then it must be by non-radiative processes. Are the planets all just still cooling off from some initial very hot state? Well Venus does in fact cool by about five degrees during its four-month-long Venus night. So if there were no sun providing energy, it appears that it could easily have cooled right down during the life of the planet. The planet Uranus shows no sign of any net loss of energy and, although it is nearly 30 times further from the Sun than Earth is, its atmosphere is far hotter at the lower altitudes.

Hopefully by now you will be starting to realise just how misleading is the propaganda about greenhouse effects on Venus or Earth, and you will want to read on to gain an understanding of what process does in fact explain all planetary atmospheric and surface temperatures, and even temperatures beneath any surface.

# 4

# Where It All Went Horribly Wrong

NASA originally produced this graphic supposedly representing how energy comes from the Sun, warming the surface and then flowing back into the atmosphere which then, at least in the troposphere, gets cooler as you go up towards the tropopause at the top, which is about 10Km to 17Km above the Earth's surface.

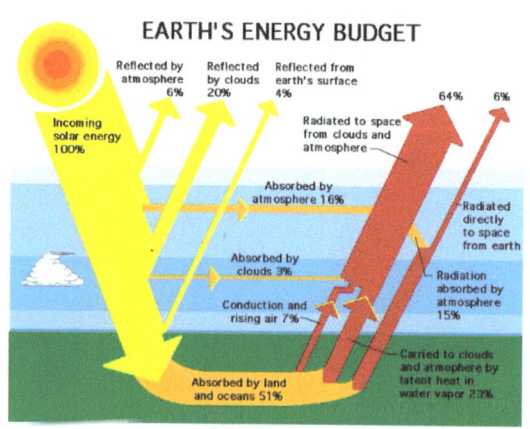

You will notice that the graphic shows 51% of the incident solar radiation penetrating the surface, where its energy is converted to thermal energy. Most of this energy (45% of the initial incoming energy) transfers from the surface to the atmosphere and the remaining 6% is radiated through the "atmospheric window" straight to space. Only one third of the 45% (namely 15%) is energy that is transferred by radiation to the atmosphere. The other two thirds of the energy from the surface that is absorbed by the atmosphere is transferred by non-radiative processes, these being 23% "by latent heat in water vapour" and 7% by "conduction and rising air." Remember that back radiation from the atmosphere can only affect the rate of cooling associated with that 15% which causes radiative cooling of the surface. Back radiation from a cold atmosphere cannot affect the rate of surface cooling that is due to the 30% of the initial energy being transferred by non-radiative processes.

We can assume that when NASA applied the SBL they then realised that the amount of energy shown as entering the surface was nowhere near enough to explain the actual observed temperatures for the surface, so you may not be able to find the above graphic on their website now. If they applied the same concepts to Venus there would be only 2% to 3% entering the surface and a far greater discrepancy between calculated and observed temperatures.

So, somewhere along the line someone got the idea that we needed to explain the "33 degree" difference as being due to back radiation, which they claimed doubles the amount of radiation going into the surface.

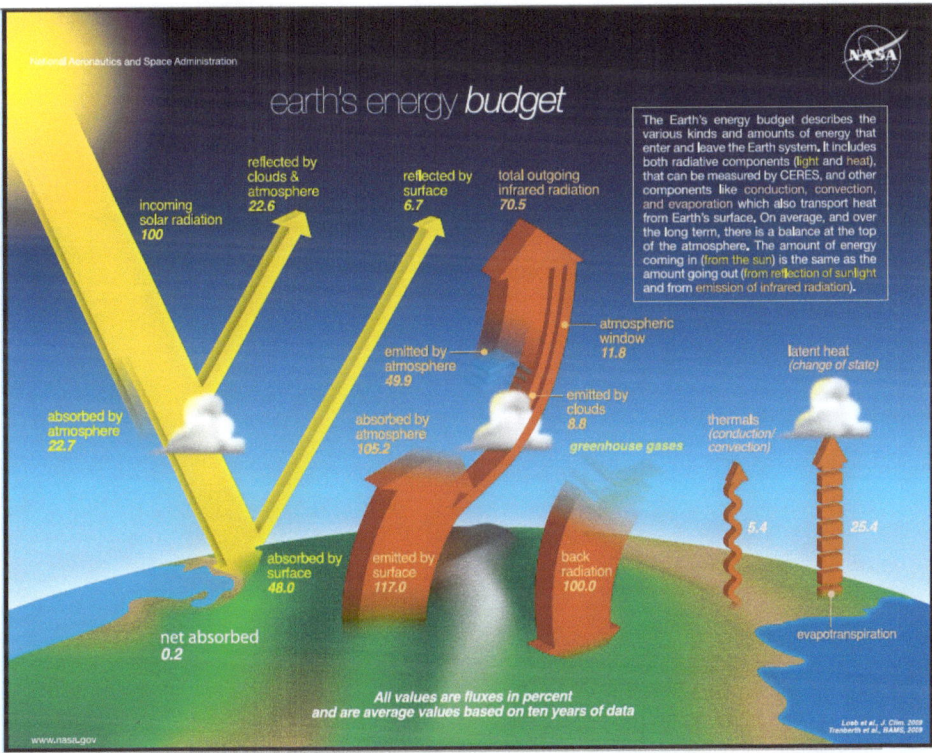

This graphic from the NASA website shows the result of adding as much back radiation as there is in all the incoming Solar radiation. We have now increased the surface radiation to 117% of the original energy that came in at the top of the atmosphere, and we have decreased the proportion of energy transferred from the surface by non-radiative processes from two-thirds right down to about one-fifth. All of a sudden, radiation has become the key player, whilst energy transferred when nitrogen and oxygen molecules collide with the surface pales into insignificance. In the lower left of the graphic we read that there is a net absorption of 0.2% supposedly causing warming. Anyone with any concept of error margins in all those figures would realise that one cannot even claim that the end result is a positive amount as it could just as easily be a net loss of 0.2% causing cooling. NASA themselves have changed the percentage absorbed by the surface from 51% in the earlier graphic to 48% in this one.

But of course the biggest fallacy in the second diagram is the fact that it clearly implies that the back radiation from a cooler atmosphere is actually transferring thermal energy to the warmer surface. As is explained in detail in the author's paper *"Radiated Energy and the Second Law of Thermodynamics"* [1 & 2] the electromagnetic energy in any such radiation is never converted to thermal energy in a warmer target such as the Earth's surface. The surface cannot be warmed to a higher temperature than that to which the Sun's direct radiation could warm it *unless* enough of the required energy gets there by some other non-radiative process. It is obvious that such a process must be occurring on other planets where temperatures are far greater, and that process will now be explained in the next two chapters.

# 5
## It's All About Restoring Equilibrium

The implication in the graphics in the previous chapter is that the surface is warmed and then energy flows into the atmosphere where warm air then rises and cools as it does so. This supposedly explains the "lapse rate" (with connotations of water running downhill) which is nothing more nor less than the observed temperature gradient in a planet's troposphere. Mind you, above the tropopause (at the top of the troposphere) temperatures then level out and start to get warmer going further up into the stratosphere. This is because ozone absorbs incident solar radiation more quickly than it can be dissipated. We see a similar effect in the thermocline just below the ocean surface, where temperatures decline rapidly because more and more Solar radiation has already been absorbed the deeper the rays penetrate.

The tropospheres of other planets also exhibit a temperature gradient, and on Venus that gradient is only a little steeper than that on Earth where it averages about 6.5 to 7 degrees per kilometre. So why do we not see hotter regions at the top of the Venus troposphere where more of the incident solar radiation is absorbed, rather like what happens in the ocean thermocline? Why does the Venus troposphere not get cooler going towards the surface where less solar radiation is received?

To help with understanding let us consider a large roughly square lake with north, south, east and west sides. In calm conditions its surface is in a state of mechanical equilibrium, apparently flat though actually shaped by gravity according to the curvature of the Earth's surface. Suppose a heavy rain storm adds new water all along the western side of the lake. We know that the additional water will be spread by gravity with much of it heading towards the eastern side. When the rain stops a new equilibrium state will eventuate with a slightly higher water level over the whole lake. If the rain had fallen all along the eastern side, there would have been water movement in the other direction, mostly towards the west, and if it had fallen in the centre of the lake the new water would have spread out in all directions. In all situations the water movement is away from the source of new water and the water level at the shoreline on all sides will have risen by an equal amount when the new equilibrium state is achieved.

We now come to the most important consideration in this book and, indeed, for the whole climate debate. We need to answer this seemingly straight forward question: *What temperature gradient, if any, should exist in the equilibrium state in a planet's troposphere?*

Most of us have been brought up with the concept that temperatures tend to "level out" in calm conditions because, as was documented in the earliest statements of the Second Law of Thermodynamics, heat transfer is supposedly only from hot to cold. Indeed we have already mentioned that any transfer of thermal energy by radiation is always from warmer to cooler bodies. Likewise, we experience this happening also by non-radiative conduction and diffusion processes, at least in a horizontal plane.

However, a physicist by the name of Josef Loschmidt postulated way back in the nineteenth century that the equilibrium state in a vertical column of any material, solid, liquid or gas would exhibit a non-zero temperature gradient (cooler at the top) due to the effect of gravitational force acting upon the individual molecules, because temperature is a measure of the mean kinetic energy (KE) of molecules and is not affected by their gravitational potential energy. However, the "heavy-weights" of the time, Maxwell and Boltzmann, rubbished the idea, and there it lay mostly buried until a handful of scientists have started to look into the idea this century. Attempts to disprove the existence of this "gravity gradient" have all been erroneous, the most common error being the overlooking of a temperature gradient also occurring in solids such as a copper wire. You cannot, for example, create a perpetual motion machine with a cylinder of gas and a wire because the wire also has a temperature gradient.

It is actually surprisingly straight forward to resolve this issue from a correct understanding of Kinetic Theory in which we consider the motion of individual molecules in the atmosphere. As you may know, such molecules in a gas are very spread out with relatively large amounts of empty space between them. They do collide, however, and have near miss "grazing collisions" and some kinetic energy will be transferred from the molecule with greater energy to the one with less. This is how thermal energy is spread out by the conduction-like process which we will call "diffusion" when a gas is involved.

If you have ever placed one of those oil filled metal column heaters in a corner of a room you will have observed this diffusion process spreading the thermal energy throughout the room. You might think this is because warm air rises and then spreads out along the ceiling and down the other side. That indeed does happen in a process called convection where we can actually observe air moving and perhaps turning a light-weight propeller. But if the heater were somehow suspended up near the ceiling then warm air would appear to move downwards. Once again, as with the lake, the

## IT'S ALL ABOUT RESTORING EQUILIBRIUM

new energy (water) moves away from the source as it endeavours to establish a new equilibrium state.

The Second Law of Thermo*dynamics* in its modern form tells us that there will be a propensity for a so-called "closed system" to tend towards a state of thermo*dynamic* equilibrium with maximum accessible entropy. The word "entropy" may be thought of as representing disorder. When there is a state of "order" then work can be done, and entropy will increase when that work is done. The greater the entropy, the less is the work that can still be done within a system. A state of "maximum accessible entropy" is one in which there are no unbalanced energy potentials and thus no means by which further work could be done within the constraints of the system. For example, if you hold an egg above the ground it may be temporarily in a state of mechanical equilibrium with an upward force from your hand balancing the downward force from gravity. Remove your hand from the system and some of the gravitational potential energy in the egg is converted to kinetic energy (as it falls) and thence partly into thermal energy and partly into energy which fragments the shell, leaving it on the floor in a somewhat greater state of disorder, but, never-the-less, a new state of equilibrium.

We can answer the question about the temperature gradient in a gravitational field if we investigate and think about just exactly what the state of thermodynamic equilibrium would be in a gas which is subjected to a gravitational field. Thermodynamic equilibrium takes into consideration all forms of energy, but the forms which are likely to change (in the absence of any chemical reaction or phase change) are kinetic energy (KE) and gravitational potential energy (PE). *The changes in such internal energy take place when molecules are in free flight between collisions.* Whilst they are, there is some interchange of KE and PE just as there is when you drop an egg to the floor.

However, the reason why an atmosphere does not just collapse to the surface is because, when a column of air is tending towards thermodynamic equilibrium the molecules that collide tend towards having the same KE at the moment of impact. This means that there is no propensity for any general air movement up, down or in any direction.

But think about what then happens to a molecule which is rising. As it does so it gains gravitational PE and loses an equivalent amount of KE, just as happens when you throw a ball upwards into the air. But if thermodynamic equilibrium is prevailing, then the next molecule it collides with at a slightly higher altitude should be expected to have an amount of KE matching the now-lower level of KE in the rising molecule. The opposite happens when molecules fall, and this is in fact the way in which a pressure gradient is maintained in a gravitational field. Pressure is proportional to the *product* of density and temperature, so we cannot assume that temperature increases merely because pressure increases. High pressure does not *maintain* high temperatures. Temperature is the independent variable in planetary tropospheres, and any given temperature can only be maintained if the supply of energy matches the loss of energy in the normal cooling processes.

Now, if there were uniform temperatures at all heights in a column of air, then the molecules at the top would have more PE but an equal amount of KE to those below them. So some would fall more than they rise and thus work would be done when they gained KE during the free fall and then increased the KE of molecules they collided with at lower heights. In short, a state of homogeneous KE at different heights in a gravitational field can never be a state of thermodynamic equilibrium with maximum accessible entropy, because work can and will be done.

So the above considerations lead to the inevitable conclusion that, at thermodynamic equilibrium, there is in fact a temperature gradient maintained by gravity because all molecules at a higher altitude have a lower mean KE (hence a cooler temperature) than those at a lower altitude. Furthermore, the difference in the mean KE should equal the difference in gravitational PE. This means that it is the sum (PE + KE) which is homogeneous at thermodynamic equilibrium, because when that is the case no further work can be done. Hence we can calculate what the temperature gradient ought to be, at least in the absence of any inter-molecular radiation, and these will be the only calculations we need in this book.

Let us consider a thought experiment in which a region of a non-radiating gas of mass $M$ all happens to move downwards by a small height difference, $H$ in a "closed system" where $g$ is the acceleration due to gravity. The loss in PE will thus be the product $M.g.H$. because a force $Mg$ moves the gas a distance $H$. But there will be a corresponding gain in KE and that will be equal to the energy required to warm the gas by a small temperature difference, $T$. This energy can be calculated using the specific heat $Cp$ and this calculation yields the product $M.Cp.T$. Bearing in mind that there was a PE loss and a KE gain, we thus have ...

$$M.Cp.T = - M.g.H$$

$$T/H = -g/Cp$$

But *T/H* is the thermal gradient, which is thus the quotient *-g/Cp*.

This result is well known, as is the fact that the atmospheres of all planets exhibit a similar temperature gradient that can be calculated from the gravitational force on that planet and the mean specific heat of the gases in its atmosphere.

However, there are small variations which reduce the magnitude of the gradient by amounts up to about a third in magnitude. This happens because of the temperature levelling effect of inter-molecular radiation which only ever transfers thermal energy from warmer to cooler regions.

So the overall true state of thermodynamic equilibrium has to take into account the propensity for non-radiative diffusion processes to produce a temperature gradient which is then reduced a little by the opposing tendency for inter-molecular radiation to level out the gradient. This is why that gradient is reduced on Earth from a theoretical -9.8 degrees per kilometre to a mean of about -6.5 to -7 degrees per kilometre, which results from radiation helping energy to "leap frog" over the slower moving diffused energy.

We are now getting close to understanding why radiating molecules have an overall cooling effect, *not* a warming effect. We have seen that they reduce the magnitude of the gradient and this causes the whole temperature plot for the troposphere to rotate about a pivoting altitude and intersect the surface at a lower "supporting" temperature. The plot must rotate like this in order to maintain radiative balance with the incoming solar radiation. When a planet's troposphere warms during its sunlit hours and cools at night, the whole temperature plot rises and falls whilst maintaining the temperature gradient. As none of this has anything to do with any "lapsing" process involving thermal energy flowing out of the surface, we will not use the term "lapse rate" when discussing the temperature gradient.. So this temperature gradient tends towards its equilibrium value simply because the Second Law of Thermodynamics tells us there will be a propensity to maintain thermodynamic equilibrium and, as we have seen, such equilibrium incorporates a temperature gradient.

# 6
# Heat Creep Provides the Answer

Recall the example of how rain falling on a part of a lake disturbs the mechanical equilibrium associated with flat, calm conditions. The extra water spreads out under the force of gravity until it covers the whole lake uniformly.

It is easy to visualise a lake "levelling out" but what happens in the troposphere where the equilibrium state is actually one in which there is a temperature gradient? We need to visualise the sloping temperature plot as if it represents the level surface of the lake, because each is an equilibrium state.

Now, if we add a "pile" of new absorbed energy somewhere in the troposphere then, like the new rain water, it also disturbs the existing equilibrium state. We can visualise it if we imagine that the sloping temperature plot is (or at least acts like) a level lake.

In the diagram below the new energy which has just been absorbed is represented by the red pile.

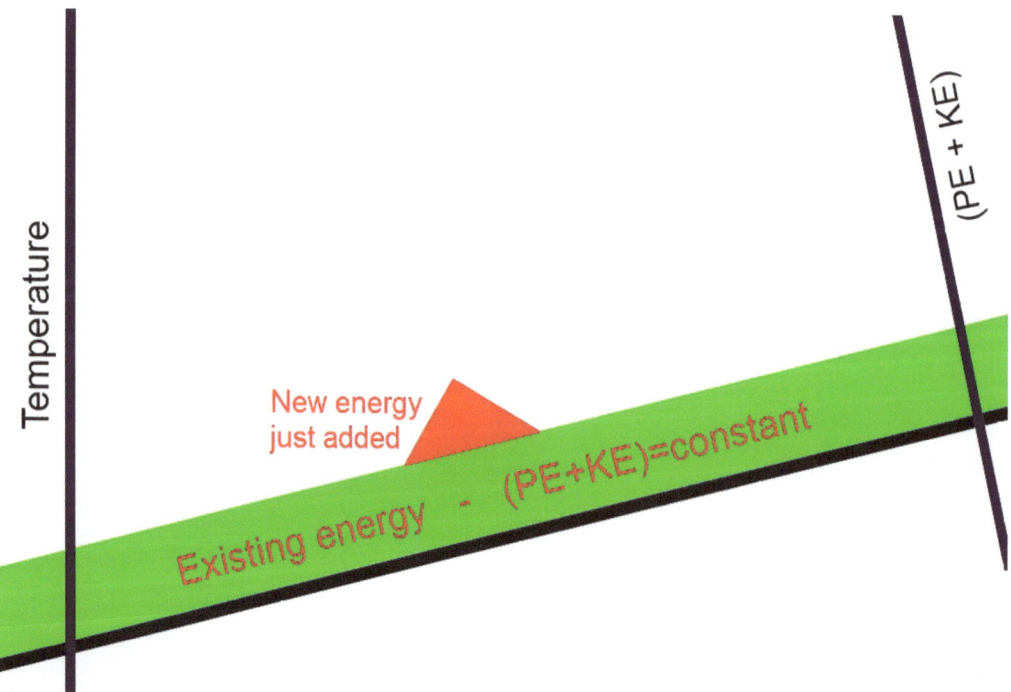

If you turn the page to the right until the (PE + KE) axis is vertical then you can imagine that red pile spreading out in all directions, so that at least some of the extra energy will in fact move slowly towards the surface and into regions that are actually warmer than those from which it came. This spreading out process starts to look like the red section in the next diagram below.

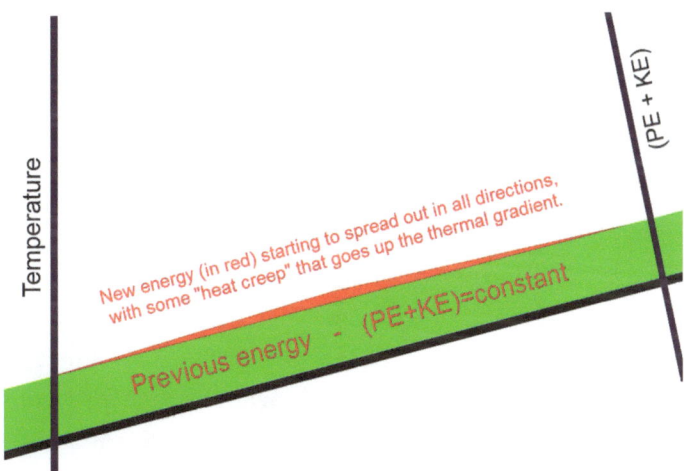

So we can now understand how the required thermal energy gets down into the lower regions of the atmospheres of planets like Venus and Uranus, even though most of the initial absorption of incident solar radiation takes place in the upper, cooler regions of their atmosphere.

*The thermal energy actually moves up the temperature gradient in a slow non-radiative process of diffusion for which we shall coin the term "heat creep" for want of a better description.*

Consider Venus in more detail now. Both its day and night time are about four Earth months long because the planet rotates far more slowly than Earth. As with any planet having an atmosphere, the overall level of the plot of temperature against altitude in the troposphere is determined by the level of incident solar radiation. This is because the whole planetary system including its atmosphere does in fact act in accord with SBL as a true black or perhaps grey body.

But what happens internally with that whole system is another matter altogether. The force of gravity sets up a temperature gradient and an associated temperature plot continues on through the surface and into the sub-surface regions where it usually becomes steeper because of the lower specific heat of the solid materials in the crust. However, the specific heat increases with temperature, so the temperature gradient in

# WHY IT'S NOT CARBON DIOXIDE AFTER ALL

Earth's mantle is far less steep than that in its outer crust. A planet's core and mantle temperatures are in fact maintained by energy from the Sun which creeps up the temperature gradient all the way to the core. There may well be other energy being created by nuclear processes, but such is not essential and is probably not sufficient. The existence of a temperature gradient in the Earth's crust does not necessarily tell us that there is a constant outflow of energy, because the gradient is just an equilibrium state. There is probably some net energy entering the crust and then being released in volcanoes and hot springs. It is doubtful that nuclear energy is created in the core of our Moon, but we should expect that core to be far hotter than the maximum temperature reached at its surface.

Let us now consider *why* the whole troposphere of Venus (together with its surface) experiences a rise in temperature of about five degrees during the daytime and an equivalent fall in the four-month night time. The reason why the temperature only falls by five degrees during the Venus night is because there is a huge amount of energy that must be radiated out of the whole troposphere in order for it all to cool by that amount. The cooling cannot be confined just to the surface, for that would upset the equilibrium and thermal energy would merely transfer back into the surface from the now hotter lowest layer of the troposphere. The temperature gradient must and will be maintained throughout the Venus troposphere while this cooling is taking place.

But when the sun rises on Venus the next morning it starts to heat the upper troposphere more than the lower troposphere, and so this disturbs the thermodynamic equilibrium. We have what is called a temperature inversion wherein, even though the temperature at the top is not hotter than that at the bottom, it is still hotter at the top than the equilibrium state would indicate it ought to be. The process of heat creep comes into play in order to attempt to restore the equilibrium, Thus the newly absorbed extra thermal energy spreads out over the troposphere away from the source, just like the new rainwater spread out over the surface of the lake.

Is the transfer of thermal energy by heat creep in violation of the Second Law of Thermodynamics because it involves heat transfer from cooler to warmer regions? The answer is a resounding "No" because the energy transfer is in fact carrying out the very process described in that law. The Second Law in its modern form says nothing at all about heat only transferring from hot to cold.

So heat creep is the non-radiative process that very obviously must be happening in order for there to be energy balance on Venus and Uranus just for starters. The vast majority of the energy on these planets has been absorbed from Solar radiation over the life of the planet. Gravity maintains the temperature gradient and thus traps the extra energy in the lower regions. The planets cannot cool off any more during their night time before they start to warm back up by an equivalent amount the next day. Their atmospheric and surface temperatures are not determined by the instantaneous absorption of Solar radiation, but by the height of the atmosphere and the temperature plot for which the gradient is determined from the $-g/C_p$ quotient (modified by inter-molecular radiation) and the overall level is that which is maintaining radiative balance.

Exactly the same process happens on Earth because physics is universal. We have discussed how the direct solar radiation reaching Earth's surface is not sufficient to warm its surface to the observed mean temperature because, as we saw in the first NASA graphic, about half of the incoming Solar radiation does not even reach the surface. In the early morning more absorption of Solar radiation takes place in the upper troposphere, which the Sun's rays strike before they reach the lower regions because of the curvature of the Earth and topography of the surface. Hotter temperatures in the stratosphere (due to ozone absorption) also cause new energy to transfer to the upper troposphere. Later in the morning the lower troposphere may absorb more and also get warmed by the surface. Wherever new energy is absorbed in the atmosphere it will then spread out in all accessible directions away from its source as it endeavours to restore thermodynamic equilibrium. The process is similar to the overall warming of the troposphere of Venus, although on Earth there will be significant extra solar energy absorbed by the surface in the absence of cloud cover. The surface cools quickly as this extra energy dissipates in the late afternoon and early evening, but when it has done so, the cooling process slows because now, as on Venus, the whole of the troposphere has to cool by an equivalent amount to the surface so as to maintain the temperature gradient. The fact that this slower cooling is readily observed provides strong support for the concept of an equilibrium state with a "gravity gradient" in the troposphere.

All of these considerations indicate that heat creep must also occur on Earth and, in effect, help to support warmer temperatures at the base of the troposphere, and thus in the surface itself. The Sun could never raise the local temperature to what is observed if the surface had not been kept warmer the night before by the interaction with the air in the lowest layer of the troposphere. Both radiative and non-radiative cooling slow down as the temperature gap narrows between the surface and that layer. This is the "supporting" mechanism we are talking about.

To summarise, there is evidence that the base of the atmosphere is supporting what are usually only slightly warmer temperatures in the surface. For the same reason that Venus only cools by five degrees at night, Earth's cooling also slows down to almost nothing in the early pre-dawn hours in calm conditions. This minimum "supporting temperature" is pre-determined by the temperature plot in the troposphere and, because of the less steep slope in moist areas, these regions are cooler. This has been verified by the study published in the Appendix. Greenhouse gases contribute to cooler surface temperatures, not warmer ones.

# 7

## So Why Is the Globe Warming?

The quick answer to this question as to what is causing Global Warming is that it all has to do with natural overlapping temperature cycles with differing time periods that are probably governed by some process relating to planetary orbits.

There appear to be two dominating natural cycles. The longer cycle has alternating periods of about 500 years of warming and cooling whilst the shorter cycle is associated with alternating periods of about 30 to 35 years of warming and about 30 years of cooling.

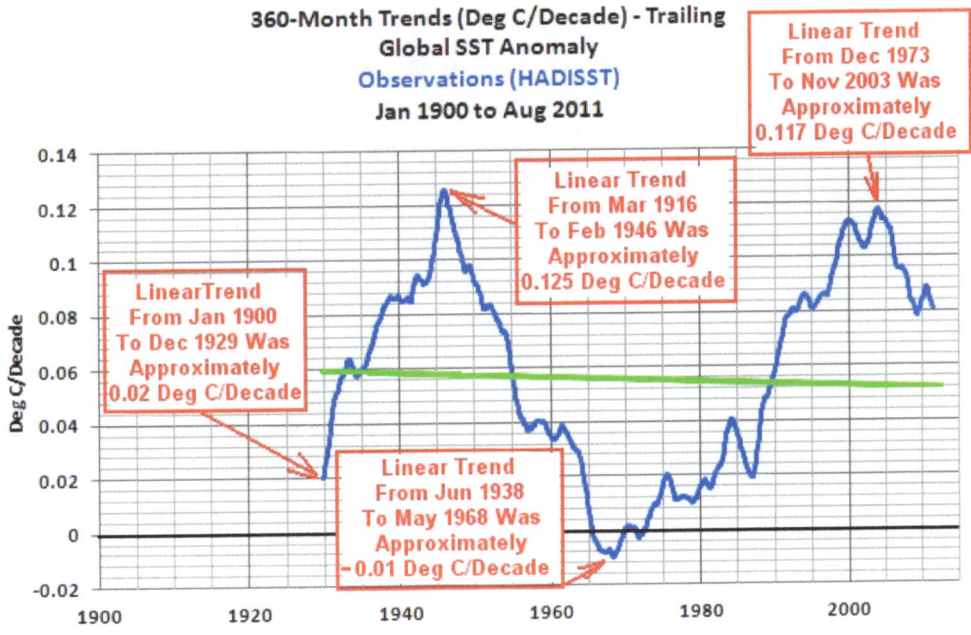

The shorter cycle takes about 60 to 65 years to complete, and we can say with reasonable certainty that it was on the rise in the 30 to 35 years from 1968 up to about 1998 to 2003, and it now appears to be in the middle of 30 years of cooling. That cooling is however partly offset by the fact that the long term (roughly 1,000 year) cycle is still rising out of the "Little Ice Age" the mid point of which was about 400 years ago. So it is not surprising that the temperatures in the 1998 to 2003 period were so hot, but the good news is that there may be just another 100 years or so in which the long term cycle may rise by only about half a degree before it then turns to about 500 years of cooling. But, even over that whole 500 years of cooling the mean temperatures are unlikely to drop by more than two degrees.

This plot from the Appendix of the author's paper [1 & 2] is interesting for two reasons. It makes the 60 to 65 year cycle quite obvious and it also shows that the mean rate of increase is actually decreasing, as indicated by the green line.

This is in keeping with the assumption that we are approaching the maximum of the 1,000 year cycle, and the rate of increase (about half a degree per century) is still about the same rate of warming that started just after the end of the Little Ice Age, not in the twentieth century.

There was very little sunspot activity during the Little Ice Age and it is thought that the level of such activity may be an indicator of Earth's climate. Whether or not there is a direct link may be hard to determine. There may be gravitational or magnetic influences from planets that affect sunspot activity. Then sunspot activity may determine cosmic rays levels, and these may affect cloud formation and thus climate on Earth. So planetary orbits probably could provide a mechanism which in some way regulates the natural climate cycles on Earth.

Whatever the case, we have seen that carbon dioxide does not cause warming and its minuscule cooling effect (probably less than a tenth of a degree) is insignificant. Thus mankind cannot expect to control Earth's climate by limiting anthropogenic emissions of carbon dioxide.

# Appendix: Temperature-Rainfall Correlation

It is a fundamental requirement for there to be a radiative greenhouse effect that water vapour and suspended water droplets in the atmosphere should have a warming effect, because these are by far the most prolific greenhouse gases. This warming effect is supposed to account for most of the "additional 33 C degrees" in surface temperatures, increasing the thermal gradient from an assumed initial isothermal (level gradient) state to one in which the surface temperature is about 30°C warmer. Then carbon dioxide and other radiating molecules are supposed to raise the temperature a little more up to a total of 33 degrees above the level gradient value. Furthermore, if carbon dioxide levels increase, it is assumed that the level of water vapour would increase as a result, and so more warming is expected, multiplying the effect of carbon dioxide with this extra positive feedback.

However, it is well known and acknowledged that water vapour leads to a lower thermal gradient, otherwise known as the "wet" or "moist" adiabatic lapse rate. Rather than the dry rate (calculated from the *-g/Cp* quotient to be -9.8C/Km) high levels of water vapour are known to reduce the gradient to about -7C/Km and even down to -6.5C/Km in the very humid Equatorial regions. The main argument in this book would thus suggest that, because water vapour makes the thermal gradient less steep, we should expect a lower surface temperature when the new radiative equilibrium is established. Thus it appears that water vapour should have a net cooling effect.

It seems remarkable that this apparent contradiction does not appear to have been investigated with what could be a relatively low cost study, compared with the funds that have been spent on other climate research. Because of this, the author spent just a few hours analysing temperature and rainfall data for 15 cities, in order to give an indication of how a more comprehensive study could be conducted.

It was considered most appropriate to select towns and cities within the tropics, which extend between the Tropic of Cancer (at about 23.5° North) to the Tropic of Capricorn (at about 23.5° South) because the Sun will be directly overhead any particular city twice a year. By selecting data for the hottest month this will usually correspond to the month in which the Sun passed through its Zenith, or the following month. As other variables may have affected the Northern Hemisphere, it was decided to limit the study to the Southern Hemisphere and to select the hottest month out of January, February or March, though nearly all turned out to be January. Such a selection avoids the need to make compensations for the angle of the Sun at latitudes outside the tropics.

It is noted that flat islands such as Singapore have very regular maximum and minimum daily temperatures, and this is almost certainly due to diffusion, convection and wind from the air just above the ocean surface, where the air temperature is governed by the water temperature. A similar effect occurs to a lesser extent with coastal cities, as well as with some cities that are close to large inland bodies of water. Hence it was decided not to include cities that were less than 100Km from the coast or such bodies of water.

It was also considered that there would be a need to adjust temperatures to what would be expected at a common altitude, and 600m was selected. Cities with altitudes outside the range 0 to 1200m were then excluded so that errors relating to assumed thermal gradients (lapse rates) would be unlikely to exceed about half a degree at the most. It was decided to use a gradient of -7C/Km for the third with the greatest rainfall, -8C/Km for the third with the least rainfall and -7.5C/Km for the middle third of the cities in the sample.

The above exclusions tend to rule out Indonesia, Papua New Guinea and other Equatorial island regions such as are found to the north of Australia. As the study was restricted to the Southern Hemisphere, it was decided to limit it to latitudes from 16.0 to 24.0 degrees south as this would include Alice Springs in Australia (latitude 23°40'S) which was considered close enough to the Tropic of Capricorn, as well as most tropical regions in Australia (AU) except those close to the Northern coastline. It also of course included a slice of both Africa (AF) and South America (SA) and, from these three continents, a total of 15 cities were selected, there being six in Australia but only four in South America where several were ruled out by altitude.

Cities which were within one degree of either the latitude or longitude of a previously selected city were not considered. However, once it was determined that a city met the requirements for altitude and coordinates, it was included in the study before referring to any temperature or rainfall data, so none were excluded for any "exceptional" reasons relating to such data, except for Emerald in Queensland Australia for which the source of data [3] had no rainfall information.

# APPENDIX: TEMPERATURE-RAINFALL CORRELATION

It is appreciated that rainfall may not be an accurate indicator of the thermal gradient, but neither would relative humidity be any better, because suspended water droplets also play a part in reducing the gradient, as does the release of latent heat when it rains.

The data is presented below in a format which the reader could use for further spreadsheet analysis:

## TABLE OF TEMPERATURE AND RAINFALL DATA FOR 15 TROPICAL CITIES

**City, Country/State, Continent, Altitude, Maximum, Minimum, Rainfall, Adj* Max, Adj Min**

01: Manaus, Brazil, SA, 39m, 27.3, 18.7, 238.7, **23.4, 14.8**

02: Goiania, Brazil, SA, 749m, 30.1, 19.5, 209.6, **31.1, 20.5**

03: Kadoma, Zimbabwe, AF, 1160m, 28.6, 17.7, 183.2, **32.5, 21.6**

04: Halls Creek, Western Australia, AU, 422m, 36.6, 24.4, 164.9, **35.4, 23.2**

05: Charters Towers, Queensland, AU, 336m, 33.5, 22.4, 164.7, **31.7, 20.6**

06: Pedro Juan Caballero, Paraguay, SA, 563m, 29.9, 20.4, 160.4, **29.6, 20.1**

07: Mariscal Jose Felix Estigarribia, Paraguay, SA, 151m, 35.4, 22.9, 129.3, **32.0, 19.5**

08: Mount Isa, Queensland, AU, 356m, 36.4, 23.7, 117.3, **34.6, 21.9**

09: Francistown, Botswana, AF, 1001m, 30.8, 18.9, 115.5, **33.8, 21.9**

10: Maun, Botswana, AF, 943m, 32.2, 19.8, 109.4, **34.8, 22.4**

11: Ghanzi, Botswana, AF, 1100m, 32.4, 19.3, 104, **36.4, 23.3**

12: Longreach, Queensland, AU, 193m, 37.1, 23.3, 73.0, **33.8, 20.0**

# WHY IT'S NOT CARBON DIOXIDE AFTER ALL

13: Beitbridge, Zimbabwe, AF, 456m, 33.5, 21.9, 56.8, **32.3, 20.7**,

14: Paraburdoo, Western Australia, AU, 389m, 41.2, 26.0, 51.4, **39.5, 24.3**

15: Alice Springs, Northern Territory, AU, 545m, 36.9, 21.8, 39.9, **36.5, 21.4**

*At 600m: for 01 to 05 use gradient 7C/Km, 06 to 10 use 7.5C/Km, 11 to 15 use 8C/Km

## Means of Adjusted Daily Maximum and Daily Minimum Temperatures

Wet (01-05):      **30.8°C**      **20.1°C**

Medium (06-10):      **33.0°C**      **21.2°C**

Dry (11-15):      **35.7°C**      **21.9°C**

APPENDIX: TEMPERATURE-RAINFALL CORRELATION

## Conclusions:

There is clearly no indication of any warming effect related to water vapour, and so no evidence for the assumed positive feedback, which is a fundamental building block for the greenhouse conjecture. Rather, the opposite appears to be the case, and water vapour does in fact appear to have the cooling effect anticipated by the hypothesis in this book.

It may well be argued that the sample was not large enough, but this must surely indicate a need for some attempt to confirm such a crucial assumption, which is vital for there to be any validity in the greenhouse conjecture that carbon dioxide has a warming effect. If water vapour does in fact have a negative feedback (as it radiates heat to higher, cooler regions, or direct to space) then so too would carbon dioxide have such a cooling effect, albeit far less in magnitude.

## Reference URL's

[1] http://principia-scientific.org/publications/psi_radiated_energy.pdf

[2] http://tallbloke.files.wordpress.com/2012/03/radiated_energy.pdf

[3] http://worldweather.wmo.int/pacific.htm

www.ingramcontent.com/pod-product-compliance
Lightning Source LLC
Chambersburg PA
CBHW051112180526
45172CB00002B/877